Field Guide to
The Oyster Fauna
of
Thailand

Somchai Bussarawit
Natural History Museum, National Science Museum, Thailand,
Technopolis, Khlong 5, Khlong Luang, Pathum Thani 12120, Thailand

Tomas Cedhagen
Section of Marine Ecology, Department of Biological Sciences,
University of Aarhus, Building 1135, Ole Worms allé 1,
DK-8000, Aarhus C, Denmark

Edited by:

Yoshihisa Shirayama
Seto Marine Biological Laboratory
Field Science Education and Research Center, Kyoto University

Kenji Torigoe
Graduate School of Education, Hiroshima University

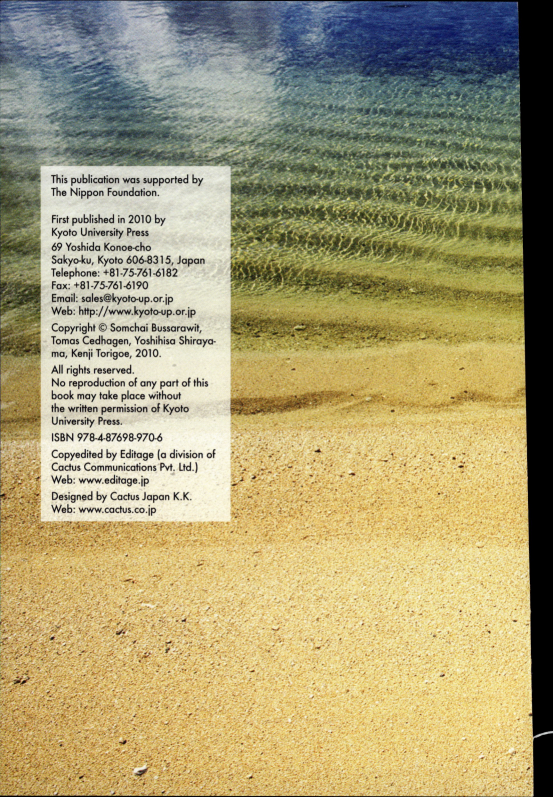

This publication was supported by
The Nippon Foundation.

First published in 2010 by
Kyoto University Press
69 Yoshida Konoe-cho
Sakyo-ku, Kyoto 606-8315, Japan
Telephone: +81-75-761-6182
Fax: +81-75-761-6190
Email: sales@kyoto-up.or.jp
Web: http://www.kyoto-up.or.jp

Copyright © Somchai Bussarawit,
Tomas Cedhagen, Yoshihisa Shirayama, Kenji Torigoe, 2010.

All rights reserved.
No reproduction of any part of this book may take place without the written permission of Kyoto University Press.

ISBN 978-4-87698-970-6

Copyedited by Editage (a division of Cactus Communications Pvt. Ltd.)
Web: www.editage.jp

Designed by Cactus Japan K.K.
Web: www.cactus.co.jp

Preface

This book has been produced to assist in implementing the NaGISA (Natural Geography In Shore Area) program (http://www.nagisa.coml.org), one of field programs participating in the project Census of Marine Life (CoML: http://www.coml.org). The purpose of NaGISA is to compare biodiversity of coastal marine ecosystems especially soft-bottom sea grass beds and hard-substratum macroalgae habitats in a global scale. In order to accomplish this goal of research program, it is prerequisite to standardize sampling and sample treatment methods. NaGISA published a handbook in 2007 from Kyoto University Press to explain these methods in detail under financial support from Nippon Foundation.

During the course of the program, NaGISA found that identification of the organisms collected through the field work constituted one of the main obstacles that needed to be overcome. As one step toward solving this problem, NaGISA has had a series of taxonomy training workshops. Further, the most up-to-date taxonomic information was described in the handbook. In addition, a large volume of taxonomic information is available through databases in the internet world. However, NaGISA recognized that a printed field guide is still very useful, since such a book might be the only information source available to those scientist living in countries that do not have rich libraries that can provide old and rare taxonomic literature. NaGISA therefore decided to publish a series of field guides to marine organisms that would be sampled during the course of its research. Although it would have been ideal to publish an encyclopedic field guide covering all taxonomic groups from around the world in a single volume, NaGISA decided to publish a series of field guides, each issue of which would cover just a single taxonomic group from one country, so as to publish vivid information promptly. Thanks to the generous financial support from the Nippon Foundation to NaGISA in the western Pacific region, the first volume in the NaGISA field guide series on the Asteroidea and Holothurioidea of Malaysia could be published in June 2008. This volume has been followed by further field guides, publication of which was again made possible by significant financial support from the Nippon Foundation. This volume does not cover all the bivalves from a country that holds great marine biodiversity. It does, however, contain descriptions of probably all the oysters found in the region—a very important group from both ecological and commercial points of view.

I hope this series of books will help all participants of NaGISA to identify their samples, and that the biodiversity database of NaGISA will be more complete and accurate based on this publication.

March 2010,
Yoshihisa Shirayama
PI, NaGISA

Contents

Abstract	5
Introduction	6
History of Thailand Oyster Study	7
Materials and Methods	9
Family Gryphaeidae	11
Family Ostreidae	14
Discussion	36
Acknowledgements	37
References	38

Abstract

Here, we describe 16 species of true oysters from Thai waters. They are widely distributed on various intertidal and subtidal substrates in the Gulf of Thailand and in the Andaman Sea. The different species were identified on the basis of their shell morphology, and their characteristic features are described. We have also included notes on their habitats and distribution. The species belonging to the following families are described in this book: family Gryphaeidae: *Hyotissa hyotis, Parahyotissa imbricata*; family Ostreidae (subfamily Crassostreinae): *Crassostrea belcheri, C. bilineata, Saccostrea cucullata, S. echinata,* and *S. forskali*; (subfamily Lophinae): *Lopha cristagalli, Dendrostrea folium, D. rosacea*, D. crenulifera*, D. sandvichensis*,* and *Anomiostrea coralliophila*;* (subfamily Ostreinae): *Nanostrea exigua*, Planostrea pestigris, Pustulostrea tuberculata*.* The 6 species indicated with asterisk were recorded for the first time in Thailand.

Introduction

The species belonging to the family Ostreidae are called true oysters. They are marine organisms, and their distribution ranges from tropical to temperate seas; most species of oysters are found in the tropical areas but do not occur in the Arctic or Antarctic zones (Harry 1985). Epifaunal molluscs are true oysters that attach to firm substrates by cementation of the left valve. Their shell shapes are generally subcircular to elongate oval, but often the shape vary depending on the type of substrate. The external shell color varies in most species, and most shells are grayish white in color with shades of dark purple in the form of irregular markings.

All oysters lack hinge teeth, but many species develop ridges or pustules along the margin of the shell; these ridges are called chomata (Stenzel 1971). Oysters are monomyarian (with a single large adductor muscle), and the foot is entirely lost in post-larval life after the larvae settle and attach to the substrate.

History of Thailand Oyster Study

Because of the variability in the shell form of oysters and the fact that some species have distinct ecomorphs, it becomes difficult to identify different species of oysters. In Thailand, oysters are widely distributed in the Gulf of Thailand and the Andaman Sea. The literature on the species of Thai oysters is confusing and taxonomical studies on these species are limited (Yoosukh and Duangdee 1999).

Lynge (1909) reported on the marine Lamellibranchiata during the Danish Expedition to Siam (1899–1900). In all, he recorded 4 species and 3 varieties belonging to the family Ostreidae, namely, *Ostrea hyotis* Linnaeus, 1758, *O. cucullata* (Born, 1780), *O. cucullata* var. *Forskali* Chemnitz 1785, *O. hyotis* var. *imbricata* Lamarck, 1819, *O. rivularis* Gould, 1861, *O. paulucciae* Crosse, 1869, and *O. cucullata* var. *Barclayana* Sowerby, 1871.

Nielsen (1976) reported 91 species of bivalves from the Phuket Marine Biological Center (PMBC), of these 91 species, 7 species were oysters: family Gryphaeidae: *Hyotissa hyotis*; family Ostreidae: *Alectryonella plicatula*, *Lopha cristagalli*, *Ostrea* sp. A, *Ostrea* sp. B, *Saccostrea cucullata*, and *S. echinata*. However, descriptions of these 7 oyster species were not provided.

Tantanasiriwong (1979) reported the following 9 species of oysters collected from the Andaman Sea coast of Thailand; family Gryphaeidae: *Hyotissa hyotis* and *H. numisma*; family Ostreidae: *Alectryonella plicatula*, *Crassostrea gigas*, *Crassostrea* sp., *Lopha cristagalli*, *L. folium*, *Saccostrea cucullata*, and *S. echinata*.

Nateewathana *et al.* (1981) recorded 10 species of oysters collected from the Andaman Sea coast of Thailand, which are deposited in the Reference Collection of PMBC, the following species were collected: family Gryphaeidae: *Hyotissa hyotis* and *H. numisma*; family Ostreidae: *Alectryonella plicatula*, *Crassostrea belcheri*, *C. gigas*, *Lopha cristagalli*, *L. folium*, *Ostrea* sp., *Saccostrea cucullata*, and *S. echinata*.

Nateewathana (1995) reported 3 species of oysters among commercial and edible mollusks of Thailand namely *Crassostrea belcheri*, *C. lugubris* and *Saccostrea cucullata*.

Yoosukh (1988) reported 12 species, 6 genera and 2 families of oysters in

Thailand including species belonging to the family Gryphaeidae: *Hyotissa hyotis* and *H. numisma* and the family Ostreidae: *Crassostrea belcheri*, *C. lugubris*, *Crassostrea* sp. 1, *Crassostrea* sp. 2, *Dendostrea* sp. I, *Dendostrea* sp. 2, *Lopha cristagalli*, *Saccostrea cucullata*, and *S. echinata*.

Yoosukh and Duangdee (1999) reported 9 species of common living oysters from Thailand which belonged to the family Gryphaeidae: *Hyotissa hyotis* and *Parahyotissa imbricata* and the family Ostreidae: *Crassostrea belcheri*, *C. iredalei*, *Dendostrea folium*, *Lopha cristagalli*, *Saccostrea cucullata*, *Saccostrea forskali*, and *Striostrea (Parastriostrea) mytiloides*.

Yoosukh (2000) reported 15 species of oysters in Thailand which belonged to the family Gryphaeidae; *Hyotissa hyotis* and *H. numisma* and family Ostreidae; *Crassostrea belcheri*, *C. lugubris*, *Saccostrea cucullata*, *S. echinata*, *S. mordax*, *Dendostrea folium*, *Dendostrea* sp., *Lopha cristagalli*, *Planostrea pestigris*, *Ostrea* sp. I, *Ostrea* sp. 2, *Saccostrea* sp. 1, and *Saccostrea* sp. 2.

Day et al. (2000) reported 3 different groups of *Saccostrea* species from Thailand on the basis of allozymes and shell morphology: *Saccostrea commercialis*, *S. cucullata*, and *S. manilai*.

Yoosukh and Sukhsangchan (2000) conducted allozyme electrophoresis in order to differentiate the genetic characters; this led to the confirmation of 3 species of tuberculate oysters from the Ranong province, Thailand namely *Striostrea (Parastriostrea) mytiloides*, *Saccostrea cucullata*, and *S. forskali*.

Klinbunga et al. (2001) conducted random amplified polymorphic DNA (RAPD) analysis of 5 species of oysters from Thailand, namely, *Crassostrea belcheri*, *C. iredalei*, *Saccostrea cucullata*, *Saccostrea forskali*, and *Striostrea (Parastriostrea) mytiloides*.

Bussarawit et al. (2006) performed phylogenetic analysis of Thai oysters and Bussarawit and Simonsen (2006a, b) studied the genetic variation in *C. belcheri*, and *C. iredalei*, and 4 species of *Saccostrea*.

The aim of this field guide is to summarize the existing knowledge on the Thai oyster fauna and to include new information on the basis of museum collections and by sampling both in the Gulf of Thailand and in the Andaman Sea and neighboring countries.

Material and Methods

The specimens used in this study were collected from the Gulf of Thailand and Andaman Sea. Collections were made from coastal areas (rocky shores and mangroves), offshore islands and oyster farms. The samples were fixed in 10 % formalin buffered with borax in seawater, usually for one week, and then preserved in 70 % ethanol. The studied specimens were deposited in the Reference Collection of PMBC.

The first author has also visited and studied the oyster species from the largest zoological museum in Copenhagen, London, Paris, and Sydney. None of the museum has collected oysters from Thailand systematically. The museum in Copenhagen included one sample from Lynge (1909), and some PMBC duplicates, previously from Nielsen (1976), were donated to the museum in London.

Oysters tightly attached to the substrates by using the left valve (LV). Their shells are commonly oval, but vary depending on the species or environments. The surface of the shell includes radial ribs, and the shells can vary from strong to weak across species. Commissural plications originating from radial ribs also vary from prominent to the obscure depending on species. A commissural shelf develops at the shell margin. Small ridgelets (minute depressions) and pits are commonly found on the commissural shelf on both sides of the ligament or encircle the entire valve, except in the genus *Crassostrea*; ridgelets known as anachomata develop on the right valve (RV) and pits known as catachomata develop on the LV. The origin of the chomata is different from those of the hinge teeth and sockets on the hinge plate of normally dentate bivalves; the chomata develop secondarily (Torigoe 1981). The plications, commissural shelf, and chomata are considered to be involved in defense against predators (Torigoe 1981). The outer shell surface bears various sculptures: growth squamae, lamellae, hyote spines, or tubular spines. The adductor muscle scar located on the inner surface of the valve, and its shape and position are unique to each species. Fingerprint-like structures are uniquely found on the shells of the genus *Alectryonella* (Torigoe 1981).

Shell characters are very important for the identification of oyster species and have been compared across species in this field book. In this study, we analyzed the shell characteristics and the soft part would be studied in the future study. The following shell characters were used to differentiate between species: general size, shape, radial ribs, commissural plication, hyote spine, chomata, commissural shelf, shape and color of adductor muscle scar, umbonal cavity, and attachment area of the LV (Torigoe 1981).

The classification was made according to Vaught 1989, with generic placement determined on the basis of the descriptions provided by Stenzel 1971, Torigoe 1981 and Harry 1985.

A few terms should be defined. Chomata (sing. choma) is a collective term for anachomata, which are small tubercles or ridgelets on the periphery of the inner surface of RV, and catachomata, which are the pits in LV for reception of anachomata; both are generally restricted to the vicinity of the hinge, but may sometimes encircle the entire valve (Stenzel 1971).

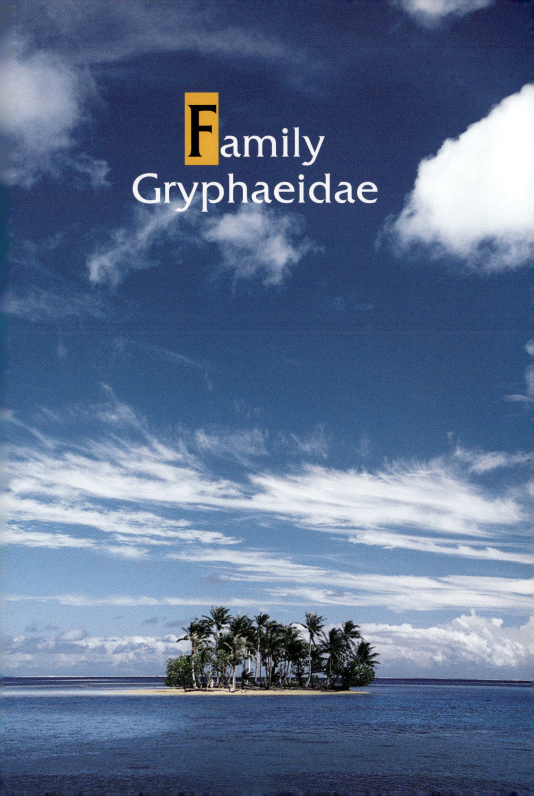

Family Gryphaeidae

Subfamily Pycnodonteinae
Genus *Hyotissa* Stenzel, 1971

Mytilus hyotis Linnaeus, 1758
Hyotissa hyotis (Linnaeus, 1758)

Common Names
Honeycomb oyster (Abbott 1974), Hyotoid oyster (Sowerby 1870, Thomson 1954).

Habitat
Attached to hard substrates, mostly in coral reef areas.

Geographical Distribution
South Africa, Indo-Pacific (Harry 1985; Yoosukh and Daungdee 1999); Amami-Oshima (Japan), and southwards (Torigoe 1981), Red Sea, Indian Ocean, Philippines, Taiwan, Hong Kong, Hainan, South China Sea to Okinawa and Honshu, Japan (Bernard et al. 1993).

Distribution in Thailand
Chumphon, Ranong, Phuket, Krabi (Yoosukh & Duangdee 1999; Yoosukh 2000), Krabi, Phuket, Satun, Ranong.

Remarks
The examined specimens of this species fit well with the species diagnosis by having large size, strong and stabilized commissural plications and thick shell. Ventral margin of the elliptical adductor muscle scar elevated. Chomata long, vermiculate, and arborescent. Lath chomata develop occasionally. Vesicular shell structure present but its quality varies from weak to strong (Torigoe 1981).

This species can be identified easily because of its strong and stabilized commissural plication, thick shell, and large size. On hard substrates, attachment area on LV is normally large.

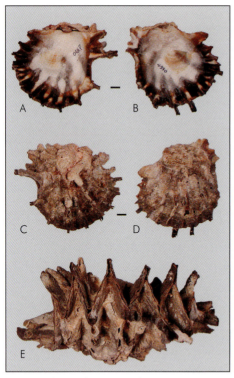

Figure 1. *Hyotissa hyotis*. PMBC no. 4330, Panwa Cape, Phuket, February 17, 1972. A. internal view of left valve; B. internal view of right valve; C. external view of left valve; D. external view of right valve; E. ventral view. Scale bars = 10 mm.

Genus *Parahyotissa* Harry, 1985

Parahyotissa macgintyi Harry, 1985
Parahyotissa imbricata (Lamarck, 1819)

Common Name
Imbricated oyster (Sowerby 1870).

Habitat
Littoral to sublittoral zone.

Geographical Distribution
Australia, Philippines, Ryukyu Island, Boso Peninsula (Japan) and southwards (Torigoe 1981) to South Asia (Yoosukh and Duangdee 1999).

Distribution in Thailand
Chon Buri, Prachuab Khiri Khan (Yoosukh & Duangdee 1999), Krabi, Chon Buri, Pattani, Rayong, Ranong, Prachuab Khiri Khan (this study).

Remarks
The examined specimens of this species fit well with the description of the genus (Harry 1985) and the species (Torigoe 1981) in being medium sized, orbicular, and equivalve. Surface with dichotomous radial ribs from the umbo in both valves. Both valves concave, but RV deeper. Ribs bear prominent growth squamae, which rise and become hyote spines. Chomata short, vermiculate. Commissural shelf large; vesicular shell structure well recognized.

This species was misidentified by Tantanasiriwong (1979) and Yoosukh (1988, 2000) as *Hyotissa numisma* (PMBC no. 1442). In *Parahyotissa numisma* (Lamarck, 1819), the shell outline is rather subcircular, and the LV is extensively cemented to the substrate. The valves are shallow with short steep sides, with the upper valve being more or less flat and sculptured.

Figure 2. *Parahyotissa imbricata*. PMBC no. 1442, Dam Khwan Island, Krabi, March 9, 1976. A. external view of right valve; B. internal view of right valve; C. external view of left valve; D. internal view of left valve. Scale bars = 10 mm.

Family Ostreidae

Rafinesque, 1815

Subfamily Crassostreinae — Torigoe, 1981
Genus *Crassostrea* — Sacco, 1897

Ostrea virginica Gmelin, 1791
Crassostrea belcheri (Sowerby, 1871)

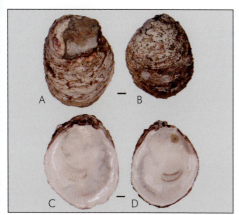

Figure 3. *Crassostrea belcheri*. PMBC no. 14264, Ang Sila, Chon Buri, February 7, 1995. A. external view of left valve; B. external view of right valve; C. internal view of left valve; D. internal view of right valve. Scale bars = 10 mm.

Common Names
Belcher's oyster (Sowerby, 1870), White scar oyster (Thailand).

Habitat
Estuarine. This species is commercially cultured in Thailand.

Geographical Distribution
Karachi, Bombay, Lake Chilka, Culcutta, Arakan, Mergui, Malaysia, Singapore, Siam Bay, Vietnam, Hong Kong, Niquebar, Sumatra Java, Sarawak (Borneo), Sula Sea, Celebes, Philippines (Ranson 1967). South China Sea, Gulf of Thailand, Andaman Sea and Indonesia (Yoosukh & Duangdee 1999). Indo-Pacific, Philippines, South China Sea, Hainan (Bernard et al. 1993).

Distribution in Thailand
This species is commercially farmed in Thailand cultured along the coast of Trat, Chanthaburi, Surat Thani, Songkla, Narathiwat, Ranong, Phangnga, Krabi, Trang (Yoosukh & Duangdee 1999), Surat Thani, Songkla (Yoosukh 2000), Surat Thani, Chon Buri, Ranong, Songkla, Krabi (this study).

Remarks
The examined specimens fit well with the descriptions of the species (Sowerby 1871; Yoosukh 1988, 2000; Yoosukh and Duangdee 1999) in being large sized, elongated orbicular, subequivalve with thick shells. Surface with numerous growth squamae on the RV. Both valves concave; LV commonly deeper than the RV. No commissural plications. Adductor muscle scar crescentic shaped, white in color. Umbonal cavity medium. No chomata. Commissural shelf narrow.

This species can be identified easily because it does not have chomata, is large in size, and has a white muscle scar.

Crassostrea bilineata
(Röding, 1798)

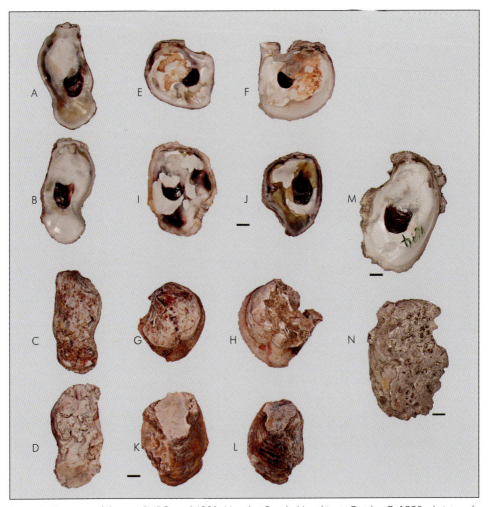

Figure 4. *Crassostrea bilineata*. PMBC no. 14281, Narathat Beach, Narathiwat, October 7, 1995. A. internal view of right valve; B. internal view of right valve; C. external view of right valve; D. external view of left valve; E. internal view of right valve; F. internal view of right valve; G. external view of right valve; H. external view of left valve; I. internal view of left valve; J. internal view of right valve; K. external view of left valve; L. external view of right valves; M. internal view of left valve; N. external view of left valve. Scale bars = 10 mm.

Common Names
Slipper-shaped oyster (Blanco et al. 1951), Black scar oyster (Thailand).

Habitat
Marine and estuarine. This species is commercially cultured in Thailand.

Geographical Distribution
South China Sea, Andaman Sea, Gulf of Thailand, Papua New Guinea (Yoosukh & Duangdee 1999). Indo-Pacific, Philippines, South China Sea (Bernard et al. 1993). India.

Distribution in Thailand
Chanthaburi, Trat, Prachuab Khiri Khan, Songkla, Nakhon Si Thamarat, Ranong, Phangnga (Yoosukh & Duangdee 1999), Samut Sakorn, Phangnga, Chon Buri, Surat Thani, Rayong, Narathiwat, Chanthaburi, Prachuab Khiri Khan, Trat.

Remarks
This species is different from *C. belcheri* in having kidney-shaped black or purple adductor muscle scar, oblong shape, flat RV, and rather foliaceous shells.

The species name, *C. iredalei*, was recently used for oysters from Thailand instead of the previously used synonym *C. lugubris* by Nateewathana et al. (1981), Yoosukh (1988, 2000), Department of Fisheries (1993), and Nateewathana (1995). The junior synonym to this species is *O. lugubris* (Sowerby, 1871) (not Conrad 1857 or Kauffman 1965).

The specific name *Ostrea lugubris* was derived from a fossil species from Cretaceous and Tertiary periods from East and Red River, Canadian, New Mexico, and Santa Fé road (Conrad 1857). The species named *Ostraea lugubris* was recorded from North America (Sowerby 1871). The specific name *Lopha lugubris* was used for a fossil recorded in the Middle and upper Turonian period from southern Colorado, New Mexico and Texas by Kauffman (1965). Hence, our opinion is that the specific names *O. lugubris* and *L. lugubris* are fossils recorded from the United States and should not be used for living species in tropical SE Asia. This fossil species must be included under a different taxon from that of *C. iredalei*.

The valid name of *Crassostrea iredalei* becomes *Crassostrea bilineata* (Röding, 1798). The reason is that this oyster was illustrated as *Ostrea parasitica indiae orientalis* by Chemnitz (1785, pl. 71, fig. 660), was named as *Ostrea edulis* var.. by Gmelin (1791). This oyster was named as *bilineata* by Röding. But Dillwyn (1817) named this oyster as *orientalis* and Küster (1868, figure was a copy of Chemnitz, pl. 8, fig. 2), Hanley (1856) and Hidalgo (1905) also used Dillwyn's name *orientalis*. Lamarck (1919) named *cristata* var. (b). Sowerby (1817) did not show any figure because of lack of specimens that correspond with *orientalis*. Anyhow they were all based on the Chemnitz's very rough figure and descriptions. Lamy (1929) stated that *bilineata* should be applied to this oyster, and because of the International Code of Zoological Nomenclature, Röding's *bilineata* becomes valid name Torigoe and Bussarawit (2010).

Genus *Saccostrea* Dollfus & Dautzenberg, 1920

Ostrea (Saccostrea) saccellus
Dujardin, 1835
Saccostrea cucullata
(Born, 1778)

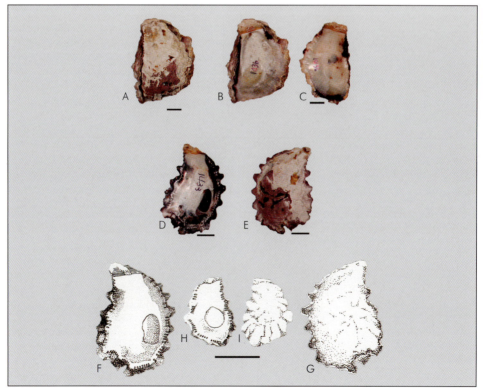

Figure 5. *Saccostrea cucullata*. PMBC no. 1436 (upper row), Tarutao Island, Satun, May 23, 1975. A. both valves; B. internal view of left valve; C. internal view of right valve; PMBC no. 1433 (middle and lower rows), Kham Yai Island, Ranong, January 19, 1975. D. internal view of right valve; E. external view of right valve; F. internal view of right valve; G. external view of right valve; H. internal view of right valve; I. external view of right valve. Scale bars A–E = 10 mm, F–G = 20 mm.

Common Names
Hood oyster (Sowerby 1871).

Habitat
Marine. Attached to hard substrates in marine environment. Found in intertidal and shallow water.

Geographical Distribution
Tropical Indo-Pacific from the Red Sea and East Africa to Australia and as far north as Japan (Yoosukh & Duangdee 1999). East Indies, Red Sea, Australia, Philippines, Moluccas, South and East China Seas, Hainan, Hong Kong, Taiwan, Indian Ocean and Red Sea (Bernard et al. 1993).

Distribution in Thailand
Chanthaburi, Trat, Chumphon, Ranong, Phuket (Yoosukh & Duangdee 1999), Phangnga, Ranong, Satun, Phuket, Chon Buri, Trang, Trat.

Remarks
The examined specimens fit well with the description of Born (1780), Morris (1985), and Yoosukh and Duangdee (1999), by having a thick shell, small size, subtriangular outline, crenulated shell margin, and straight and short hinge line. LV usually attached and deep cupped with a distinct umbonal cavity. Dorsal posterior side raised vertically, ventral and anterior sides flat. RV rather flat. Adductor muscle scar elliptical, white or stained with purple. Chomata well developed, rod shaped, may extend partially or completely around the valves. Commissural shelf not developed. Attachment area large.

This species is validly referred to as *Saccostrea cucullata* (Born, 1778). Although Born's original description of this species records the spelling as "*cuccullata*" the name as printed is divided between the double c's justifying the conclusion that this original incorrect spelling was indeed as inadvertent error (ICZN Art. 32 (a) (ii)). Born obviously was aware of this error and emended the spelling to "cucullata" in the next edition of his work (Born 1780) (see also ICZN Art. 33 (a)(i)) (Morris 1985).

This species has many synonyms, especially in the tropic. *S. cucullata* is widely distributed throughout the tropical Indo-Pacific. The type locality recorded from Ascension Island, and Atlantic Ocean was questioned as an error of locality recorded by many authors.

We found an error in the species name referred to in fig. 34b and fig. 34c in Sowerby (1871). Sowerby called *S. cucullata* as *Lopha cristagalli*.

We consider *Saccostrea amasa* (Iredale 1939; Anan 1959; Thomson 1954; Carroen 1969), *S. mordax* (Sowerby 1871; Kira 1959, 1962; Habe & Okutani 1975; Habe 1977, 1981; Torigoe 1981; Yoosukh 2000), *S. mordax* ecomorph *sueli* (Oliver 1992), *Ostrea forskali* var. *mordax* (Lamy 1929) as junior synonyms of *S. cucullata*.

Saccostrea echinata
(Quoy et Gaimard, 1835)

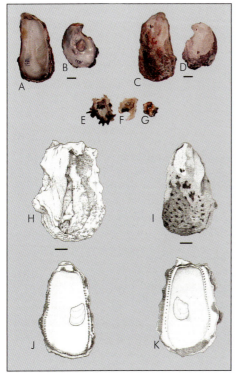

Figure 6. *Saccostrea echinata*. PMBC no. 1431, Panwa Cape, Phuket, July 14, 1974. A. internal view of right valve; B. internal view of right valve; C. external view of right valve; D. external view of right valve; E. external view of right valve; F. external view of left valve; G. internal view of right valve; H. external view of left valve; I. external view of right valve; J. internal view of right valve; K. internal view of left valve. Scale bars = 10 mm.

Common Names
Black bordered oyster (Sowerby 1870), Spinous oyster (Sowerby 1870), Black edge oyster (Thomson 1954).

Habitat
Marine. Attached to hard intertidal and subtidal substrates.

Geographical Distribution
Australia; Barrier Reef, Thursday Island, Indonesia, Philippines (Thomson 1954). Ishigaki Island, Japan and southwards (Torigoe 1981).

Distribution in Thailand
Phuket (Yoosukh 2000), Phuket, Ranong, Trang, Krabi.

Remarks
The examined specimens fit well with the description of Torigoe (1981), by having large size, outline elongate oval, inequivalves. LV deeply concave, shell margin rising particulary high at ventral margin. RV flat, but usually with a hump between hinge and lip, ventral margin rising particularly in large specimens, bearing dense lamellar growth squamae at the shell margin. Commissural shelf wide, black or purple in color. Chomata strong and elongate, encircle the valves. Umbonal cavity deep. Attachment area large.

Young specimens often have black tubular spines rising vertically on the shell surface of the RV.

This species was misidentified by Nielsen (1976) as *S. cuccullata* according to the photograph of the species and deposited specimens in the PMBC reference collection (PMBC no. 1430,1431). We consider *S. spinosa* (Deshayes 1836; Iredale 1939; Sowerby 1871) as a synonym of *S. echinata*.

Saccostrea forskali
(Gmelin, 1791)

Figure 7. *Saccostrea forskali*. PMBC no. 8446, Narathat Beach, Narathiwat, December 3, 1991. A. external view of left valve; B. external view of right valve; C. internal view of left valve; D. internal view of right valve; E. external view of right valve; F. internal view of right valve; G. external view of left valve; H. internal view of left valve. Scale bars = 10 mm.

Common Name
Indian rock oyster (Nagabhusnam & Maine 1991), Bombay oyster (Awati & Rai 1931).

Habitat
Attached to hard substrates in marine to estuarine areas where it expresses different ecomorphs. This species is commercially cultured in Thailand.

Geographical Distribution
South China Sea, Andaman Sea (Yoosukh & Duangdee 1999).

Distribution in Thailand
This species is commercially farmed in Thailand in Trat, Chanthaburi, Chon Buri, Prachuab Khiri Khan, Nakorn Sri Thammarat, Songkla, Ranong, Krabi, Phangnga, Trang, Satun (Yoosukh & Duangdee 1999), Narathiwat, Samut Sakorn, Rayong, Chon Buri, Trat, Chanthaburi, Prachuab Khiri Khan, Ranong, Songkla, Phangnga.

Remarks
This commercial species fits well with the description provided by Lamy (1925, 1929) and Yoosukh and Duangdee (1999). It has a small-to-medium sized shell, variable in shape, shell margin irregularly crenulated. LV cup shaped, with distinct umbonal cavity. RV soft, overlap LV, having thin lamellae at periphery. Adductor muscle scar kidney shaped, with alternate strips of white and brown.

For a long time, this species was misidentified in Thailand as *Saccostrea cucullata* (Jarayabhand et al. 1994; Department of Fisheries 1993; Nateewathana 1995; Yoosukh 1988). It was recorded as *S. cucullata* var. *forskali* in the Gulf of Thailand by Lynge (1909). Allozyme studies have confirmed that *S. forskali* and *S. cucullata* are different species (Yoosukh & Sukhsangchan 2000; Bussarawit 2006b).

Day et al. (2000) reported *S. commercialis* from Thailand. They found it to be identical with the samples from Australia. Their finding are debatable, because the information on morphology is very sparse, and no plates were published. The species identified as *S. commercialis* in their papers might probably be *S. forskali*, but this needs to be further clarified. In contrast, by using RAPD, Klinbunga et al. (2001) found *S. commercialis* from Australia to be clearly different from all other Thai *Saccostrea* spp. Our own study performed on the basis of DNA sequencing also supports the view that *S. commercialis* does not occur in Thailand (Bussarawit et al. 2006). We also found *S. forskali* to be clearly distinct from Australian *S. commercialis* samples (Bussarawit & Simonsen 2006b).

Subfamily Ostreinae   Rafinesque, 1815
Genus *Nanostrea*   Harry, 1985

Ostrea deformis
Lamarck, 1819
Nanostrea exigua
Harry, 1985

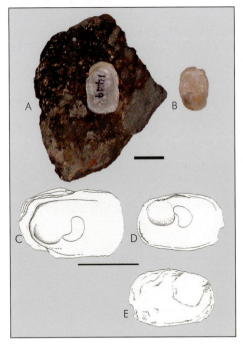

Common Name
None.

Habitat
Attached to rock and empty shells, intertidal and shallow areas.

Geographical Distribution
Madagascar, Red Sea, India, Malaysia, Indonesia, Philippines, Japan (Ranson 1967). Tropical Indo-Pacific, Red Sea to Hawaii (Harry 1985). Indo-Pacific, Indian Ocean, Red Sea, South and East China Seas to Okinawa and Hawaii (Bernard et al. 1993).

Distribution in Thailand
Phuket.

Remarks
The specimens of this species fit well with the description provided by Harry (1985) and Lamprell and Healy (1998), in being of very small size, outer surface of RV may not be lamellose. Large attachment area on LV, which has the upturned part plicated. Chomata prominent, often extending around the margin.

This species was recorded for the first time in Thai waters.

Figure 8. *Nanostrea exigua*. PMBC no. 1449, PMBC Beach, Phuket, March 20, 1976. A. internal view of left valve; B. internal view of right valve; C. internal view of left valve; D. internal view of right valve; E. external view of right valve. Scale bars = 10 mm.

Genus *Planostrea*   Harry, 1985

Ostrea pestigris
Hanley, 1846
Planostrea pestigris
(Hanley, 1846)

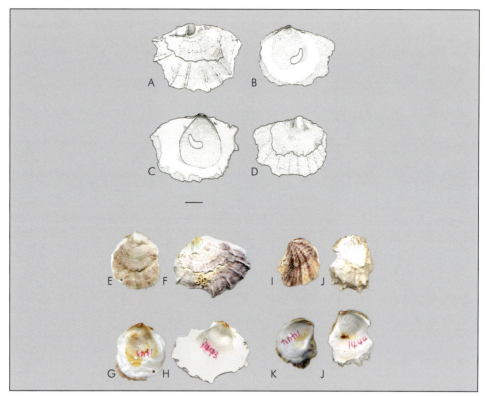

Figure 9. *Planostrea pestigris*. PMBC no. 1443, Dam Khwan Island, Krabi, March 18, 1975. A. external view of left valve; B. internal view of right valve; C. internal view of left valve; D. external view of right valve; E. external view of right valve; F. external view of left valve; G. internal view of right valve; H. internal view of left valve; PMBC no. 1444 Yao yai Island, Phangnga. March 17, 1975; I. external view of right valve; J. external view of left valve; K. internal view of left valve. Scale bars = 10 mm.

Common Names
Cat's foot oyster, Palm-footed oyster (Sowerby 1871).

Habitat
Coastal area, often emerges at low tide.

Geographical Distribution
Indo-West Pacific, Philippines, Formosa, Thailand, North Borneo, Australia (Harry 1985). Philippines, South and East China Sea, Hong Kong, Taiwan, Yellow Sea, Shandong to Honshu, Japan (Bernard et al. 1993).

Distribution in Thailand
Chon Buri (Yoosukh 2000), Phangnga, Krabi, Phuket.

Remarks
The examined specimens fit well with the description provided by Morris (1985) and Yoosukh (2000), having small size, being subquadratic in outline and very compressed. Background color overpatterned by radial streaks of dark purple, interior white. Commissural shelf wide, especially on LV. Chomata small, extending short distance anterior and posterior to the hinge. Adductor muscle scar relatively small and elongated, usually white at the interior shell. No umbonal cavity.

This species was misidentified by Tantanasiriwong (1979) as *Hyotissa numisma* (PMBC no. 1443, 1444). *Crassostrea* sp. 2 of Yoosukh (1988) should be identified as *Planostrea pestigris*.

We consider *Ostrea rivularis* (Gould 1861; Ranson 1967), *Ostrea paulucciae* (Crosse 1869, 1870; Hidalgo 1905; Lamy 1929; Ranson 1960), *Ostrea palmipes* (Sowerby 1871; Talevera & Faustino 1933; Blanco et al. 1951), *Ostrea (Crassostrea) paulucciae* (Tchang & Lou 1956), *Dendostrea paulucciae* (Habe 1977), as synonyms of *P. pestigris*.

Genus *Pustulostrea*   Harry, 1985

Ostrea tuberculata Lamarck, 1804
Pustulostrea tuberculata (Lamarck, 1804)

Common Names
Tuberculated oyster (Sowerby, 1870).

Habitat
Marine. Found in shallow water and usually associated with coral reefs.

Geographical Distribution
Andaman Island, Singapore, Indonesia, Sharks Bay (Australia) (Ranson 1967). New Hebrides Island (Harry 1985). Philippines, Indonesia, South China Sea, Hainan (Bernard et al. 1993). Formosa – Amami-Oshima (Japan) (Inaba and Torigoe, 2004).

Distribution in Thailand
Phuket.

Remarks
The examined specimens fit well with the diagnosis of Harry (1985), having a medium to large size, compressed, elongated dorso-ventrally. Beak of LV is disproportionately large, often being half the height of the shell. Attachment area small. LV exterior without plications, but usually having low round pustules, closely but irregularly spaced, variably produced on the surface. RV flat, subcircular with its beak slightly protruding: outer shell layer covered with closely spaced, smooth, fragile, appressed lamellae margin. Hinge exceptionally long, umbonal cavity deep. Adductor muscle scar usually dark brown on LV and white on RV. Chomata ostreine, extending at least to the dorsal margin.

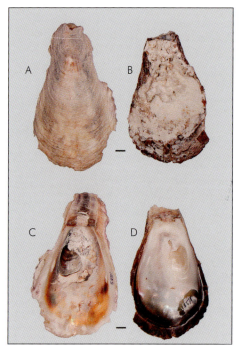

Figure 10. *Pustulostrea tuberculata*. PMBC no. 1425, Mai Thon Island, Phuket, May 5, 1976. A. external view of left valve; B. external view of right valve; C. internal view of left valve; D. internal view of right valve. Scale bars = 10 mm.

This species was misidentified by Tantanasiriwong (1979) as *C. gigas* (PMBC no. 3673) and as *H. numisma* (PMBC no. 1440). This species is easily distinguished from other species by the pustulose external shell surface and a long hinge with a deep umbo cavity. *Ostrea* sp. (Yoosukh 1988; 2000) should be identified as *P. tuberculata*.

This species was recorded for the first time in Thai waters.

Subfamily Lophinae Vyalov, 1936
Genus *Lopha* Röding, 1798

Mytilus cristagalli
Linnaeus, 1758
Lopha cristagalli
(Linnaeus, 1758)

Common Names
Coxcomb (cock's comb) oyster (Sowerby, 1871).

Habitat
Attached to rocks and corals by clasping spines of the LV; subtidal. They normally grow on gorgonians.

Geographical Distribution
Kii peninsula (Japan) and southwards through Indo-Pacific (Torigoe 1981). Red Sea, India, Indonesia, Philippines, Japan, Australia (Harry 1985). Indo-Pacific (Yoosukh and Duangdee 1999). Indo-Pacific, Red Sea, Celebes Sea, South China Sea, Taiwan (Bernard *et al.* 1993).

Distribution in Thailand
Phuket, Krabi, Chumphon (Yoosukh & Duangdee 1999), Chumphon, Phuket (Yoosukh 2000), Phuket.

Remarks
The examined specimens fit well with the diagnosis of Torigoe (1981), having sharp radial ribs. Commissural plications irregular and zigzag shaped. Surface of both valves, with small low round protuberances, pustules. Commissural shelf medium. Anachomata small, grow on both valves, scattered over the shell margin in 2-4 irregular rows. Adductor muscle scar half-moon shaped.

This species is easily distinguished from other species by having a sharp zigzag margin and pustulate shell surface.

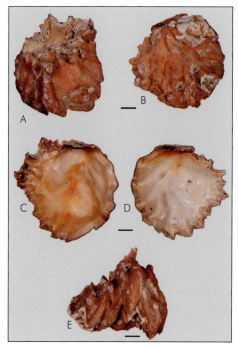

Figure 11. *Lopha cristagali*. PMBC no. 10082, Mai Thon Island, Phuket. July 20, 1993. A. external view of left valve; B. external view of right valve; C. internal view of left valve; D. internal view of right valve; E. ventral view. Scale bars = 10 mm.

Genus *Dendrostrea*   Swainson, 1835

Ostrea folium
Linnaeus, 1758
Dendrostrea folium
(Linnaeus, 1758)

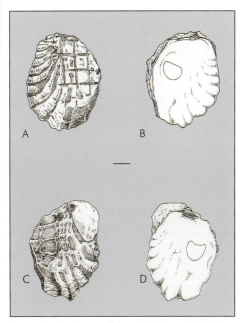

Figure 12. *Dendrostrea folium*. PMBC no. 17670, Tung Woa Lan, Chumphon. February 5, 2000. A. external view of left valve; B. internal view of left valve; C. external view of right valve; D. internal view of right valve. Scale bars = 10 mm.

Common Names
Leaf oyster (Sowerby 1871), Tree oyster (Born, 1780), Bronze oyster (Thomson 1954).

Habitat
Attached to rocks, dead shells, and cage nets in marine and estuarine areas.

Geographical Distribution
Kii Peninsula (Japan) and southwards to Indo-Pacific (Torigoe 1981). Indo-Pacific (Yoosukh & Duangdee 1999). Indo-Pacific, Australia, Philippines, South and East China Seas, Hong Kong, Hainan, Taiwan to Honshu, Japan (Bernard *et al.* 1993).

Distribution in Thailand
Chon Buri, Chanthaburi (Yoosukh & Duangdee 1999), Chon Buri, Rayong (Yoosukh 2000), Phuket, Krabi, Ranong, Pattani, Prachuab Khiri Khan, Chumphon, Chon Buri.

Remarks
The examined specimens fit well with the description of Torigoe (1981), and Yoosukh and Duangdee (1999), having a small to medium size, thin, elongate oval; margin irregularly folded. Both valves concave having and have dichotomous radial ribs from umbo, top of ribs round. Both valves with many fine and imbricate growth squamae, which are sometimes eroded dorsally. Adductor muscle scar kidney shaped. Chomata few. Commissural shelf narrow. Umbonal cavity shallow.

Torigoe (1981) considered Iredale's (1939) so-called *D. brescia* type on flat substrates and Stenzel's (1971) *Lopha folium* ecomorph 2 attached on the stem of gorgonians as identical with *D. folium*. Abbott (1974) stated that *Ostrea frons* became oval and resembled *D. brescia* on reefs, but remained long and oval and hence resembled *D. folium* on gorgonian stems.

Stenzel (1971) considered *O. frons* to be a synonym of *O. folium*. However, Oliver (1992) preferred to use *D. frons* as a valid name for the species in the Indo-Pacific and *D. folium* as the valid name for the species in the West Indian region. Our opinion is that the name *D. folium* should be used for the tropical Indo-Pacific species, particularly as the type locality is Ambon (Indonesia).

We consider *Ostrea (Pretostrea) bresia* (Iredale 1939), *Lopha folium* ecomorph 2 (Stenzel 1971), and *Lopha cristagalli* (Springteen & Leobrera 1986) as synonyms of *D. folium*.

Dendrostrea rosacea
(Deshayes, 1836)

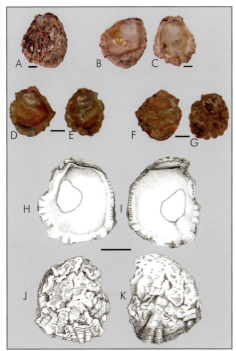

Figure 13. *Dendrostrea rosacea*. PMBC no. 10922, Ao Yon, Phuket. April 20, 1997. A. both valves; B. internal view of left valve; C. internal view of right valve; D. internal view of left valve; E. internal view of right valve; F. external view of left valve; G. external view of right valve; H. internal view of left valve; I. internal view of right valve; J. external view of left valve; K. external view of right valve. Scale bars A–G = 10 mm, H–K = 20 mm.

Common Names
Rosy oyster (Sowerby 1871).

Habitat
Marine. Attached to hard substrates, subtidal.

Geographical Distribution
Sagami Bay and south to South-West Pacific (Torigoe 1981). Philippines, South China Sea, Taiwan, Okinawa to Honshu, Japan (Bernard et al. 1993).

Distribution in Thailand
Phuket, Ranong.

Remarks
The examined specimens fit well with the description of Torigoe (1981), by their small to medium size, orbicular outline, subequivalve. Attachment area large. Both valves have dichotomous ribs from umbo, top of ribs slightly rounded, with slightly elongated growth squamae. Both valves concave. Commissural plications well crenulated. Both valves have many closely set short growth squamae. Commissural shelf narrow to medium. Chomata rarely encircle the ventral margin. Adductor muscle scar half-moon shaped. Umbonal cavity medium.

Tantanasiriwong (1979) misidentified this species as *Hyotissa numisma* (PMBC 1441).

This species was recorded for the first time in Thai waters.

Dendrostrea crenulifera
(Sowerby, 1871)

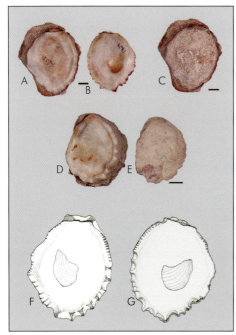

Figure 14. *Dendrostrea crenulifera*. PMBC no. 1442, Yao Yai Island, Phangnga. March 17, 1975.
A. internal view of left valve; B. internal view of right valve; C. both valves; D. internal view of left valve; E. external view of right valve; F. internal view of left valve; G. internal view of right valve.
Scale bars A–E = 10 mm, F–G = 5 mm.

Common Names
Crenuliferous oyster (Sowerby 1871), Green oyster (Thomson 1954).

Habitat
Marine. Attached to hard substrates; subtidal.

Geographical Distribution
Boso Peninsula (Japan) and southwards through Indo-Pacific (Torigoe 1981). Port Jackson (Sydney), New Hebrides, Indonesia, Vietnam, Philippines, Japan (Thomson 1954), Red Sea (Oliver 1992). Indo-Pacific, Red Sea, Philippines, South and East China Sea, Hainan, Taiwan to Honshu, Japan (Bernard et al. 1993)

Distribution in Thailand
Phangnga, Satun, Phuket, Krabi, Prachuab Khiri Khan.

Remarks
The examined specimens fit well with the description provided by Torigoe (1981), by having large and half-moon shaped adductor muscle scars. Slightly elongated small chomata encircle the entire valve or disappear on the ventral part; shell margin constantly crenulated. Attachment area large and can span even the entire LV. Commissural plications sharply crenulated. Commissural shelf narrow. Chomata as small elongate tubercles. Adductor muscle scar half-moon shaped. Umbonal cavity medium.

This species was recorded for the first time in Thai waters.

Dendrostrea sandvichensis
(Sowerby, 1871)

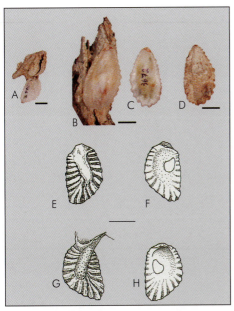

Figure 15. *Dendrostrea sandvichensis*. PMBC no. 3672, Panwa Cape, Phuket. July 14, 1994. A. both valves on top, internal view of left valve below; B. internal view of left valve; C. internal view of right valve; D. external view of right valve; E. external view of right valve; F. internal view of right valve; G. external view of left valve; H. internal view of left valve. Scale bars = 10 mm.

Common Names

Sandwich Island oyster (Sowerby 1871).

Habitat

Marine. Attached to hard substrates; subtidal.

Geographical Distribution

Originally described from Sandwich Islands, Hawaii (Sowerby 1871), Red Sea – New Hebrides – Indonesia – Vietnam – Philippines – South China – Japan (Inaba and Torigoe, 2004).

Distribution in Thailand

Phuket, Samut Sakorn, Prachuab Khiri Khan.

Remarks

The specimen is similar to the one illustrated by Sowerby (1871) and Lamprell & Helay (1998), being small sized, solid, oblong trigonal-shaped, sculpture on both valves are numerous, strong, rounded radial folds, which become sharp V-shaped folds at margins. Adductor muscle scar large, kidney-shaped. Colour white and with some green or purple patches internally.

This species was misidentified by Tantanasiriwong (1979) as *Lopha cristagalli* (PMBC nos. 1445, 1446, 1447). Pustulated shell surface is pronounced in *Lopha cristagalli*.

This species was recorded for the first time in Thai waters.

Genus *Anomiostrea* Habe & Kosuge, 1966

Ostrea pyxidata Adams & Reeve, 1850

Anomiostrea coralliophila Habe, 1975

Figure 16. *Anomiostrea coralliophila*. PMBC no. 14493, PMBC Beach, Phuket. November 13, 1994. A. external view of right valve; B. internal view of right valve; C. external view of left valve; D. external view of right valve; E. internal view of left valve; F. internal view of right valve; G. external view of right valve; H. external view of left valve; I. external view of left valve; J. internal view of right valve; K. internal view of left valve; L. internal view of left valve; M. external view of right valve; N. internal view of right valve; O. both valves. Scale bars = 10 mm.

Common Names
Box-shaped oyster (Sowerby 1871).

Habitat
Attached to hard substrates in intertidal to subtidal marine areas.

Geographical Distribution
Ceylon, Borneo, Moluccas, Java, Philippines (Ranson, 1967), Indo-West Pacific, Madagascar to the Philippines (Harry 1985). Indian Ocean, Philippines, Sumatra, South China Sea, Beibu Gulf, Guangsi (Bernard et al. 1993).

Okinawa (Japan) (Inaba and Torigoe, 2004).

Distribution in Thailand
Phuket.

Remarks
The examined specimens, as well as the description of the genus, fit well with the description of Habe (1975) and Harry (1985), by having its small size, thin valves, and subcircular outline. LV evenly swollen, nearly hemispherical, RV flat. The specimen is white tan outside and pure white within. LV attached to a flat area; attachment area being nearly at right angle to the median plane of the shell. Many small subacute radial ribs at the attachment area, separated by spaces about as wide as their diameters. Adductor muscle scars white and large, appear to be subcircular with a flattened dorsal margin, but the latter is indistinct. Ventral margin of the scars in both the valves, prominently elevated above the level of the shell surface; the space of the scar partially covered by chambering.

This species was renamed by Habe (1975) for *Ostrea pyxidata* Adams & Reeve (1848), which is the type species of the genus *Anomiostrea* Habe & Kosuke (1966). The name was preoccupied by *Ostrea pyxidata* Born, 1780, which is a different taxon.

This species collected from the PMBC beach was named *Ostrea* sp. A by Nielsen (1976). This species has distinct characters with respect to the adductor muscle scars and LV and RV shell pattern.

This species was recorded for the first time in Thai waters.

Discussion

Some oysters are commonly, or even predominantly, observed in their habitats; therefore, it is of great importance to determine their taxonomy in order to understand their roles in coastal zone ecology. This study was conducted using the material obtained from natural resources as well as that obtained from aquaculture farms. This study has contributed to clarifying the problems posed in the taxonomic classification of oysters in Thai waters.

Previous studies on oysters (families Gryphaeidae and Ostreidae) in Thailand are scarce and there is considerable confusion regarding the scientific names. The inconsistent taxonomic classification renders most of the published information uncertain. In some cases, unidentified or misidentified oysters (Nielsen 1976; Tantanasiriwong 1979) were preserved in the PMBC collection so that it was possible to revise them. Those materials have been included in this study.

The oyster fauna of Thailand now comprises 16 species belonging to 10 genera and 2 families. Of these 16 species, the following 6 are reported here for the first time: *Anomiostrea coralliophila*, *Dendrostrea crenulifera*, *D. rosacea*, *D. sandvichensis*, *Nanostrea exigua* and *Pustulostrea tuberculata*.

Some oyster species are important in aquaculture, and it is important to know their taxonomy for various reasons. It is important to select the correct species, i.e. species with high productivity, for maintaining a broodstock in an optimal way so that hybridization or other detrimental genetic effects are avoided. There are 2 species of *Crassostrea* that are commercially cultured. The identity of *C. belcheri* could be confirmed, but the other species, previously known under the name *C. lugubris* and *C. iredalei* was found to have the valid name *C. bilineata*. Another commercial species was previously known as *Saccostrea cucullata* but was correctly identified as *S. forskali*. It is an estuarine species. However, *S. cucullata* is a separate true marine species living in open sea areas and not utilized in aquaculture in Thailand. The identities of these species were determined by electrophoresis and DNA sequencing (Bussarawit *et al.* 2006).

The present study classified the species only on the basis of shell characters of adult oysters. However, the large number of individuals in the material has allowed us to sort out ecomorphological variation from such characters of importance for identification and systematics. Ecomorphological variation occurs in most oyster species, and greatest variation was found within the genus *Saccostrea*. The shell shape variation in this genus is enormous, and there is a possibility that also other species belonging to this genus can be identified in the future. Additional information on the systematics can be obtained in the future by studying the anatomy of the soft part of the species, as well as by using molecular biology techniques, e.g., DNA sequencing.

Day *et al.* (2000) reported *Saccostrea manilai* Buroker, 1979 from Thailand. However, their report is controversial, because the information on morphology is very limited and no photos were published. We speculate that it is likely that this species could be *S. echinata*.

In a parallel study, we investigated the larvae of Thai oysters. In plankton samples, we could find the larvae of *Ostrea futamiensis*, but surprisingly, the adult species could not be found.

Acknowledgements

We would like to thank the Phuket Marine Biological Center (PMBC) and DANIDA for supporting this study through the Scientific Cooperation Program; Assoc. Prof. Wantana Yoosukh, Kasetsart University, Bangkok, Thailand, Prof. Kenji Torigoe, Graduate School of Education, Hiroshima University, Japan and Prof. Yoshihisa Shirayama, Seto Marine Biological Laboratory, Kyoto University, Japan for invaluable comment to the manuscript. Mr. Pathairat Singdam for the scientific illustration, Mr. Sahate Utsaha for field assistance.

This publication was partially supported by Nippon Foundation.

References

Abbott, T. 1974. American Seashells (2nd ed.). 633 p. Van Nostrand Reihold Co., New York.

Abbott, T. 1991. Seashells of Southeast Asia. 145 p., 52 pls. Graham Brash, Singapore.

Abbott, T. and P. Dance, 1982. Compendium of Seashells. 411 p. E.P. Dutton Inc., New York.

Adams, H. and A. Adams, 1858. The genera of recent mollusca; Oysters. Vol. 2, pp. 568–569. Vol. 3, pl. 129. John Van Voost, London.

Adams, A. and L.A. Reeve, 1848. Mollusca in: The zoology of the voyage of H.M. Samarang under the command of Captain Sir E. Becher during 1843–1846. Part 4. Reeve and Benham, London.

Ahmed, M. 1971. Oyster species of West Pakistan. Pakistan J. Zool. 3: 229–236.

Ahmed, M. 1975. Speciation in living oysters. Adv. Mar. Biol. 13: 357–397.

Allan, J. 1959. Australian shells. Revised edition. 487 p., 44 pls. Oysters: 270–276. fig. 65, pls. 29, 30. Georgian House Pvt. Ltd., Melbourne.

Angell, C.L. 1986. The biology and culture of tropical Oyster. ICLARM Studies and Reviews 13, Manila, 42 p. (ICLARM Contribution No. 315). International Center for Living Aquatic Resources Management, Manila.

Arakawa, K.Y. 1990. Commercial important species of oysters in the world. Mar. Behav. Physiol. 17: 1–13.

Awati, P.N. and H.S. Rai, 1931. *Ostrea cucullata* (The Bombay oyster). Indian Zool. Mem. 3: 1–107.

Blanco, G.L., D.K. Villaluz and H.R. Montalban, 1951. The cultivation and biology of oyster at Bacor Bay, Luzon. Philippines J. Fish. 1: 45–67.

Bernard, F.R., Y.Y. Cai and B. Morton, 1993. Catalogue of the living marine bivalve molluscs of China. 146 p. Hong Kong University Press, Hong Kong.

Born, L. 1778. Index rerum naturalium musei Caesarei Vindobonensis. Part I, Testacea. 458 p. Officina Krausiana, Vienna.

Born, L. 1780. Tesacea Musei Caesarei Vindobonensis quae jussu Mariae Teresiae Augustae desposuit et descripsit. 422 p. Joannus Paulus Kraus, Vienna.

Brohmanonda, P., K. Muturasin, T. Chingpeepien and S. Amornjaruchit, 1988. Oyster culture in Thailand. In: E.W. McCoy and T. Chongpeepien (eds.). Bivalve Molluscs Culture Research in Thailand. ICLARM Technical Reports 19: 31–39.

Bussarawit, S., P. Gravlund, H. Glenner and A.R. Rasmussen, 2006. Phylogenetic analysis of Thai oysters (Ostreidae) based on partial sequences of the mitochondrial 16S rDNA gene. Phuket Marine Biological Center Research Bulletin 67: 1–9.

Bussarawit, S. and V. Simonsen. 2006a. Genetic variation in populations of white scar (*Crassostrea belcheri*) and black scar oysters (*Crassostrea iredalei*) along the coast of Thailand by means of isoenzymes. Phuket Marine Biological Center Research Bulletin 67: 11–21.

Bussarawit, S. and V. Simonsen. 2006b. Genetic variation in populations of four species of *Saccostrea* from Thailand, Malaysia and Australia measured by means of isoenzymes. Phuket Marine Biological Center Research Bulletin 67: 23–37.

Carreon, J.A. 1969. The malacology of Philippines oysters of the genus *Crassostrea* and a review of their shell characters. Proc. Nat. Shellfish Assoc. 59: 104–155.

Cernohorsky, W.O. 1978. Tropical Pacific Marine Shells. 352 p. Pacific Publication Pvt. Ltd., Sydney.

Chemnitz, H. 1785. Neues systematiches Conchylien-Cabinet. Vol. 8. Bauer and Raspe, Nürnberg.

Chen, S., Y. Wang, J. Sun, Z. Qi, X. Ma and Q. Zhuang, 1980. Studies on the Mollusca fauna of Nanji Islands, East China Sea. Acta Zool. Sinica 26(2): 171–177.

Conrad, T.A. 1857. Description of Cretaceous and Tertiary fossils. U.S. & Mexican Boundary Service Reports by Emory 1(2): 141–174, fig. 5a, 5b.

Crosse, J.C.H. 1869. Diagnoses Molluscorum Novae Caladoniae incolarum. Jour. de Conchyl. 17: 183–188.

Crosse, J.C.H. 1870. Description d'especes nouvelles. Jour. de Conchyl. 18: 97–109, 2 pls.

Dall, W.H. 1898. Contribution to the Tertiary fauna of Florida. Wagner Free Inst. Sci. 3(4): 571–947, pl. 23–35.

Dall, W.H., P. Bartsch and H.A Rehder, 1938. A manual of the recent and fossil marine pelycypod mollusks of the Hawaiian Islands. B.P. Bishop Museum Bull. 153: 233 p., 58 pls.

Day, A.J., A.J.S. Hawskin and P. Visootiviseth, 2000. The use of allozyme and shell morphology to distinguish among sympatric species of the rock oyster *Saccostrea* in Thailand. Aquaculture 187: 51–72.

Department of Fisheries, 1993. Manual for cultivation of oysters. Department of Fisheries, Ministry of Agriculture and Cooperatives. 48 p.

Deshayes, G.P. 1831. Encyclopedie Methodique ou par ordre de matieres. Histoire naturelle des Vers et Mollusques: Oysters. Paris. Vol. II: 257–594.

Deshayes, G.P. 1836. Lamarck's Histoire Naturelle des Animaux san Vertebres Oysters. Vol. 7.

Dillwyn, L.W. 1817. A descriptive catalogue of recent shells. Vol. 2, 1092 p.

Dunker, W. 1882. Index Molluscorum Maris Japonici 301 p. 16 pls. Cassellis Cattorum. T. Fisher, London.

Faustino, L.A., 1932. Recent and fossil shells from the Philippine Island I. Philippines J. Sci. 49(4): 543–549.

Galtsoff, P.S. 1964. The American oyster *Crassostrea viginica* Gmelin. Fish. Bull. Fish. Wld. Serv. U.S. 64: 1–480.

Gmelin, J.F. 1791. Caroli a Linné Systema Naturae per Regna Tria Naturae editio decimatertia (13) acuta reformata. Leipzig, 1(6), Vermes Testacea: 3021–3910.

Gould, A.A. 1850. Shells collected by the United States Exploring Expedition. Proc. Boston Soc. Nat. Hist. 3: 343–348.

Gould, A.A. 1861. Descriptions of shells collected in the North Pacific Exploring Expedition. Proc. Boston Soc. Nat. Hist. 8: 14–40.

Habe, T. 1951. Genera of Japanese shells, Pelecypoda and Scaphopoda. Ostreacea pp. 91–95, figs. 186–192.

Habe, T. 1961. Coloured illustrations of the shells of Japan II. Hoikusha, Osaka.

Habe, T. 1964. Shells of the western Pacific in color IL. 233 p, 66 pls. Hoikusya, Osaka.

Habe, T. 1975. New name for *Anomiostrea pyxidata* (Adams and Reeve) (Ostreidae). Venus 33: 184.

Habe, T. 1977. Systematics of Mollusca in Japan, Bivalvia and Scaphopoda. Hokuryu kan Publishing, Tokyo. pp. 106–111.

Habe, T. 1981. A catalogue of molluscs of Wakayama Prefecture, the province of Kii. I. Bivalvia, Scaphopoda and Cephalopoda. Ostreina: 81–84. Seto Mar. Biol. Lab., Shiyahama, Wakayama Pref.

Habe, T. and S. Kosuge, 1966a. Shells of the World in color II. 194 p., 68 pls. Hoikusha, Osaka.

Habe, T. and S. Kosuge, 1966b. New genera and species of the tropical and subtropical Pacific Molluscs. Venus 24(4): 312–341, pl 29.

Habe, T. and S. Kosuge, 1967. Common shells of Japan in colour. 223 p, 64 pls. Hoikusha, Osaka.

Habe, T. and S. Kosuge, 1979. Shells in the world in Color II. 194 p. Hoikusha Publishing, Osaka.

Habe, T. and T. Okutani, 1975. The Molluscs of Japan II. Gakken Illustrated Nature Encyclopedia. Gakken, Tokyo.

Hanley, S. 1846. A description of new species of *Ostrea*, in the collection of H. Cuming. Proc. Zool. Soc. London. 13: 105–107.

Hanley, S. 1856. An illustrated and descriptive catalogue of recent bivalved shells. 392 p. *Ostrea*. Wood's Index Testaceologicus. New edition.

Harry, H.W. 1981. Nominal species of living oyster proposed during the last fifty years. The Veliger 24(1): 39–45.

Harry, H.W. 1985. Synopsis of the supraspecific classification of living oysters. The Veliger 28(2): 121–158.

Hedley, C. 1910. The marine fauna of Queensland. Rep. Aust. Ass. Advanc Sci. 12: 329–371.

Hidalgo, J.G. 1905. Catalogo de los Molluscos testaceous de las Islas Filipinas. Jolo y Marianas. Rev. Real. Acad. Cienc. Exactas. Fis. Nat. Madrid. Vol. 3, 408 p.

Hirase, S. 1930. On the classification of Japanese oyster. Jap. J. Zool. 3: 1–65, 95 figs.

Hirase, S. 1932. Some more species of Japanese oysters. Jap. Jour. Zool. 4: 213–222, 4 figs.

Hirase, S. 1934. A collection of Japanese Shells. 130 p., 217 pls., Matumurasanshodo, Tokyo.

Hirase, Y. 1915. The illustration of a thousand shells. Vol. II-III.

Hornell, 1921. The common molluscus of south India. Report IV of 1921. Madras Fish. Bull. 14: 97–215.

Inaba, A. and K. Torigoe. 2004. Systematic Description of the Recent Oysters. Oysters in the World Part 2. Bulletin of the Nishinomiya Shell Museum No.3. 63 p., 13 pls.

Iredale, T. 1936. Australian molluscan notes. No. 2. Rec. Aust. Mus. 19(5): 267–340.

Iredale, T. 1939. Mollusca Part I. Ostreiformes. In Scientific Report of Great Barrier Reef Expedition, 1928–1929. British Mus. Nat. Hist. Sci. Rep. 5(6): 391–403, pl.7.

Jarayabhand, P., S. Jaraeontia, C. Srisaard and P. Menasveta, 1994. Experiments on larviculture of Thai oyster species. Thailand. Journal of Aquatic Science I (1): 43–53.

Kauffman, E.G. 1965. Middle and Late Turonian oyster of *Lopha lugubris* group. Smithsonian Misc. Coll. 148(6): 92 p., 18 figs., 8 pls.

Kay, E.A. 1979. Hawaiian Marine Shells. B.P. Bishop Mus. Spec. Publ. 64(4): 654 p. Bishop Museum Press, Honolulu.

Keen, A.M. 1971. Marine Shells of Tropical West America (2nd ed.). 624 p. Stanford University Press, California.

Kira, T. 1959. Coloured illustration of the shells of Japan I. 240 p., 71 pls. Hoikusha, Osaka.

Kira, T. 1962. Shells of the Western Pacific in color L. 224 p., 72 pls. Hoikusha, Osaka.

Kirtisinghe, P. 1978. Sea shells of Sri Lanka. 202 p., 61 pls. Charles E. Tuttle Co., Inc., Rutland.

Klinbunga, S., P. Ampayup, A. Tassanakajon, P. Jarayabhand, and Yoosukh, 2001. Genetic Diversity and Molecular Markers of Cupped Oyster (Genera *Crassostrea*, *Saccostrea* and *Striostrea*) in Thailand Revealed by Random Amplified Polymorphic DNA Analysis. Marine Biotechnology 3: 133–144.

Kuroda, T. 1930. Catalogue of shell bearing molluscs of Japan. Venus 2(3): 47–54.

Kuroda, T. 1960. A Catalogue of Molluscan fauna of the Okinawa Islands. 106 p. Ryukyu Univ., Naha.

Küster, H. C.1846. Systematisches Conchylien-Cabinet von Martini und Chemnitz. Bd. 7(1). pp. 67–84; pl. 8–15. Bauer and Raspe, Nürnberg.

Küster, H.C. and W. Kobelt, 1868. Systematisches Conchylien-cabinet von Martini und Chemnitz. pp. 67–84. Bauer and Raspe, Nürnberg.

Lamarck, J.B.M. de E. 1804. Une nouvelle espece de Trigonie, et sur un novelle espece d'huitre, decouvertes dans le voyage du capitaine Baudin. Ann. Mus. Natn. Natur. Paris. 4: 351–359, pl. 67.

Lamarck, J.B.M. de E. 1806. Memoires sur les fossiles des environs de Paris. Ann. Mus. Natn. d'Hist. Natur. Paris. 8: 156–166.

Lamarck, J.B.M. de E. 1818–1819. Histoire Naturelle des Animux sans Vertebres. Vol. 6, 258 p.

Lamprell, K. and J. Healy, 1998. Bivalves of Australia. Vol. 2. pp. 67–84. Backhuys publishers, Leiden.

Lamy, E. 1924. Notes sur les especes lamarckiennes d'*Ostrea*. Bull. Mus. Natn. d'Hist. Natur. Paris. 30: 92–99, 151–158, 231–238, 316–320.

Lamy, E. 1925. Les Huitres de la Mer Rouge (d'apres les materiausx recueillis par le Dr. Jousseaume). Bull. Mus. Natn. d'Hist. Natur. Paris. 31: 190–196, 252–257, 317–322.

Lamy, E. 1929–1930. Revision des *Ostrea* vivants du Museum National d'Histoire Naturelle de Paris. Jour. Conchyliologie. 73: 1–46, 71–108, 133–168, 233–275.

Lamy, E. 1936. Huitre De l'Indochine, Bull du Museum ser. 2, 8(5): 427–434.

Lamy, E. 1938. Huitre De l'Indochine, Bull du Museum ser. 2, 10(3): 287–291.

Linnaeus, C. 1758. Systema naturae per regna tria naturae, secundum classes, ordines, genera, species, cum characteribus, differentiis, synonymis, locis. Editio decima, reformata. (ed. 10). Vol. 1. 824 p. Laurentius Salvius, Stockholm.

Lynge, H. 1909. Marine Lamellibranchiata. The Danish Expedition to Siam 1899–1900. D. Kgl. Danske Vidensk. Selsk. Skrifter, 7. Raekke, Naturvidenskab og mathematik Afd. V. 3: 99–299.

Matsukuma, A., T. Okutani and T. Habe, 1991. World Seashells of Rarity and Beauty. 206 p., 156 pls. National Science Museum, Tokyo.

McLean, RA. 1941. The oyster of the western Atlantic. Notulae Naturae. Acad. Natur. Sci. Philadelphia. 67: 1–14, 4 pls.

Morris, S. 1985. Preliminary guide to the oysters of Hong Kong. Asian Marine Biology. 2: 119–138.

Nagabhusnam R. and U.H. Maine, 1991. Oysters in India. In: W. Menzel (ed.). "Estuarine and Marine Bivalve Mollusc Culture." pp. 201–209. CRC Press Inc., Boca Raton, Florida.

Nateewathana, A. 1995. Taxonomic account of commercial and edible molluscs, excluding cephalopods of Thailand. Phuket Mar. Biol. Center. Spec. Publ. 15: 93–116.

Nateewathana, A., P. Tantichodok, S. Bussarawit and R. Sirivejabandhu, 1981. Marine organisms in the Reference Collection. Phuket Mar. Biol. Center Res. Bull. 28: 43–86.

Nielsen, C. 1976. An illustrated checklist of bivalves from PMBC beach with a reef flat at Phuket, Thailand. Phuket Mar. Biol. Center. Res. Bull. 9: 1–7.

Oliver, P.G. 1992. Bivalves Seashell of the Red Sea. National Museum of Wales, Cardiff. 330 p.

Oliver, P.G. 1995. Bivalves (Bivalvia). In: S.P. Dance (ed.). Sea shells of Eastern Arabia. pp. 196–284. Motivate Publishing, London.

Poutier, J.M. 1998. Bivalves. In: K.E. Carpenter and V.H. Niem (eds.). The living marine resources of the Western Central Pacific. Vol. 1. pp. 124–362. FAO, Rome.

Quoy J.R T. and P. Gaimard, 1835. Voyage de decouvertes de l' Astrolabe execute par ordre de Roi, pendant les annes 1826-1829, sous le commandement de M. J. Dumont d'Urville. Zoologie, Vol. 3, Part. 2, pp. 367–954.

Ranson, G., 1948. Prodissoconque set classification des Ostreides vivants. Bull. Mus. Roy. Hist. Nat. Belgique. 24: 1–12.

Ranson, G., 1949. Note sur trois especes Lamarckiennes d'Ostéidés. Bull. Mus. Hist. Nat. Paris. 21: 248–254.

Ranson, G. 1960. Les prodissoconques (conquilles larvaires) des ostreides vivants. Bull. Inst. Oceanogr. Monaco. I: 1–41.

Ranson, G. 1967. Les especes d'huitres vivants acuellement dans le monde, defines par leurs coquilles larvaires ou prodissoconques. Etude des collections de quelques-uns des grands musees d'histoire naturene. Revue des Travaux de l'Institut des Peches Maritimes. 31: 127–199.

Röding, P.F. 1798. Museum Boltenianum sive catalogus cimeliorum e tribus regnis naturae. Vol. 2. v + 352 p., 56 text-fig. George & Cie (Geneva, Basel).

Rosell, N.C. 1991. The slipper-shape oyster (*Crassostrea iredalei*) in the Philippines. In: W. Menzel (ed.). "Estuarine and marine bivalve mollusc culture." pp. 307–313. CRC Press Inc., Boca Raton, Florida.

Sacco, F. 1897. Pelecypoda (Ostreidae, Anomiidae et Dimyidae). In: Bellardi L. and F. Sacco (eds.). I molluschi dei terreni Terziarii del Piemonte e della Liguria. Pt. 23, 66 p., 11 pl. Carlo Clausen, Torino.

Satyamurti, S.T. 1956. The mollusca of Krusadai Island II. Scaphopoda, Pelecypoda and Cephalopoda. Bull. Madras Gov. Mus. N. ser. Nat. Hist. Sect. 1, no. 2, Pt. 7: 66–69. Pl. 10–11.

Savielle-Kent, W. 1891. Oyster and oyster fisheries of Queensland. Queensland Govt. Rep. Government printer, Brisbane.

Savielle-Kent, W. 1893. The Great Barrier Reef of Australia. Its Products and Potentialities. 387 p. W.H. Allen and Co. Ltd., London.

Sommer, C., W. Schneider and J.M. Poutiers, 1996. The Living Marine Resources of Somalia. p. 376. FAO, Rome.

Sowerby, G.B. Jr. 1839. A Conchological Manual (1st Ed.). 130 p. G.B. Sowerby, London.

Sowerby, G.B. Jr. 1870–1871. Monograph of the genus *Ostraea*. In: L.A. Reeve's Conchologia Iconica. L. Reeve and Co., London. Vol. 18.

Springteen, F.J. and F.M. Leobrera, 1986. Shells of the Philippines. 277 p. Carfel Seashell Museum, Manila.

Stenzel, H.B. 1971. Oysters. In: R.C. Moore (ed.). Treatise on Invertebrate Palaeontology. Part N, 3: 953–1224. Geological Soc. America Inc. and Univ. Kansas.

Swainson, W. 1835. The elements of modern conchology: briefly and plainly stated for the use of students and travellers. 62 p. London.

Takatsuki, S. 1949. Oysters. 269 p. Gihodo, Tokyo.

Talevera F. and L. Faustino, 1933. Edible molluscs of Manila. Philippines Jour. Sci. 50(1): 1–48, 18 pls.

Tchang, Si and Tze-kong Lou, 1956. A study on Chinese Oysters. Acta Zool. Sinica 8(1): 65–97. 3 figs. 5 pls.

Tantanasiriwong, R. 1979. A checklist of marine bivalves from Phuket Island, adjacent mainland and offshore islands, Western Peninsular Thailand. Phuket Mar. Biol. Center Res. Bull. 27: 1–15.

Thomson, J.M. 1954. The genera of oysters and Australian species. Australian J. Mar. Freshwat. Res. 5: 132–168.

Torigoe, K. 1981. Oysters of Japan. J. Sci. of Hiroshima Univ. ser. B. Div. 1 (Zool.) 29(2): 291–419.

Torigoe K. and A. Inaba, 1981. On the scientific name of Japanese spiny oyster "kekaki." Venus 40(3): 126–134.

Torigoe, K. and S. Bussarawit, 2010. Validation of the commercial cultured black scar oyster species name, *Crassostrea bilineata* rather than *Crassostrea lugubris* and *Crassostrea iredalei* in Indo-Pacific region. Paper presented at the World Congress on Malacology 2010, Phuket Royal City Hotel, Phuket Thailand, 19–24 July 2010.

Vaught, K.C. 1989. A classification of the living Mollusca. American Malacologists Inc. Florida.

Yoosukh, W. 1988. Oysters of Thailand. p. 16. Department of Marine Sciences, Kasetsart University. (in Thai).

Yoosukh, W. 2000. Taxonomic account of oysters in Thailand. pp. 81–86. Mollusc Research in Asia.

Yoosukh, W. and T. Duangdee, 1999. Living oysters in Thailand. Phuket Mar. Biol. Center Spec. Publ. 19(2): 363–370.

Yoosukh, W. and C. Sukhsangchan, 2000. Electrophoresis studies on tuberculated oysters from Ranong province, Thailand. Phuket Mar. Biol. Center Spec. Publ. 21(2): 477–481.